THE SOLAR SYSTEM

TEACHING COMPANION

by **Kneko Burney**
Tickle+Tickle Publishing • Scottsdale, Arizona

Written by Kneko Burney
With contributions by Regina R Woodard

Graphic layout by Adriana Patricia De La Roche
And Zoe Williams Sticka
The illustrations were electronically rendered in Illustrator.
The text was set in Clearface Gothic LT Std.
The display type was set in PeachyKeenJF.
Composed in the United States of America.

Printed in the United States of America
Text copyright ©2017 by Kneko Burney

For more information, write to Tickle+Tickle Publishing,
a division of Change3 Enterprises LLC, Scottsdale, AZ.
www.discoveryrocket.com

First Edition
Library of Congress Cataloging-in-Publication Data
Burney, Kneko
Rikki & the Rocket Twins Discover the Solar System
/ by Kneko Burney p. cm.
Summary: Rikki learns about the Solar System with
the help of the Rocket Twins
ISBN-13: 978-1546745341
ISBN-10: 1546745343
[1. Science.] I.

SATURN

JUPITER

MARS

EARTH

VENUS

MERCURY

Continue the Adventure on www.discoveryrocket.com

10 9 8 7 6 5 4 3 2 1 (sc)

SUN

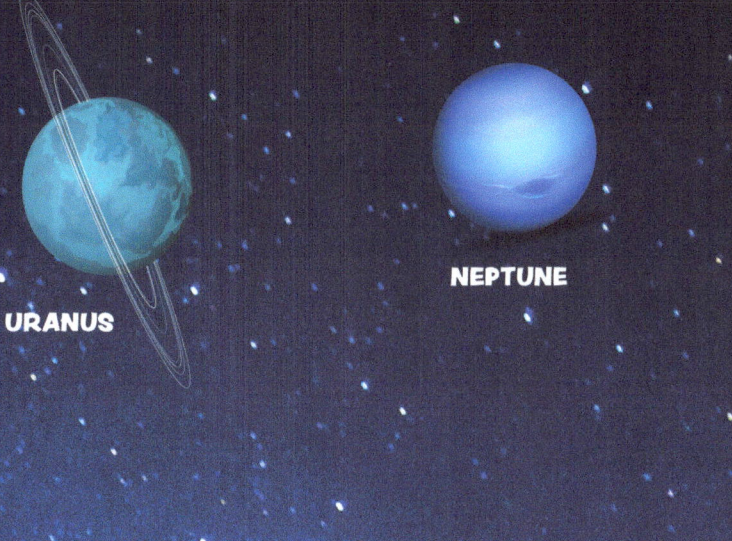

URANUS

NEPTUNE

PLUTO

ERIS

This book is dedicated to the pair
that makes my heart and mind soar…
Rikki Grace, my daughter, and
her father, Ricky Wallace.
—Kneko

This curious little girl is Rikki.

Every time she looks up at the sky, her mind is filled with questions of why?

Why is the day light and the night dark?
Why are there so many stars?
Why can't I fly beyond the clouds?

Mommy and Rikki play the Pretend Game every night before bed.

They sit on the window seat in Rikki's room and tell stories.

Rikki asks,
"Tonight can we pretend to fly my
Discovery Rocket to outer space?"

"Yes, we can!" says Mommy with
a sneaky smile on her face.

Because outer space is such a big place, Mommy wakes up the Rocket Twins to help.

The colorful pair spring to life and awaken Rikki's imagination.

Mommy tells a story about the planets in the Solar System, while Tikki in blue and Timbo in red make pretend planets dance above their heads.

They all giggle and jiggle with delight...

Even though they are having lots of fun, it is time for Rikki to get into bed.

She snuggles with still too many questions in her head.

Mommy whispers,
"It's time to discover something
amazing, my curious girl."

It is time for Rikki and the Rocket
Twins to go on an adventure that
is out of this world!

Before Rikki knows it, the three are in outer space!
Jump in your Discovery Rocket so we can race.

Race to the stars and away we go…

Up, up & up to discover the solar system.

SUN

At the center of everything in our solar system is the Sun.

Timbo quickly bakes his favorite snack in the Sun's amazing heat.

Timbo: A solar system is made up of planets floating around a star, like our Sun. The way planets float around the star is called their "orbit". Most planets also spin like a top while they orbit. The spin of a planet is called its "rotation".

The Sun is a star whose light and heat reach very, very far. It is so big and hot that all the water on Earth could not cool it down.

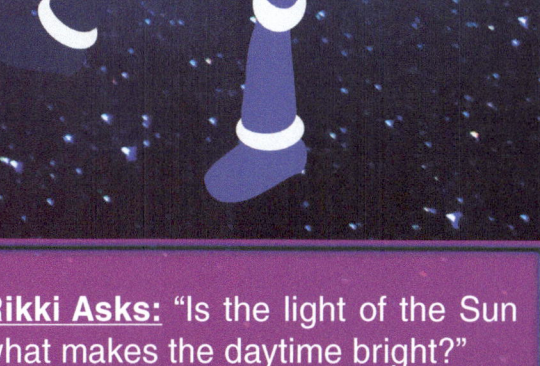

Rikki Asks: "Is the light of the Sun what makes the daytime bright?"

Yes! The Sun's light also reflects off the moon at night.

 THE SOLAR SYSTEM: The Sun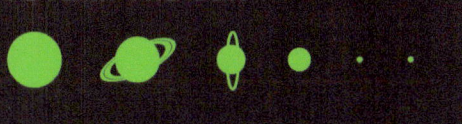

The first planet, and the one that is closest to the Sun, is Mercury.

Mercury has very little air surrounding it. That is why it gets so hot and cold down on its surface.

Tikki: The air and gases around a planet form the atmosphere. Each planet's atmosphere is made up of different gases, like oxygen and carbon dioxide, which are attracted to the planet by a force called gravity. These gases can help trap heat on the planet.

Timbo is sweating because it is very hot on the burnt, sunny side of Mercury, while Rikki and Tikki are shivering on the frozen, shadowy side.

MERCURY

Rikki Asks: "Could people live on Mercury?"

No. We need more air to breathe and it is much too hot during the day and too cold during the night.

THE SOLAR SYSTEM: Planet Mercury

The second planet is Venus. We call it the Lava Planet because it is covered by so many volcanoes!

Venus has hundreds of volcanoes, more than any other planet in our solar system.

VENUS

Timbo: Venus has a thin atmosphere, just like Earth does, but on Venus the air is mostly CO_2 (carbon dioxide). CO_2 is called a greenhouse gas because it helps trap the Sun's heat, just like a greenhouse. This is why Venus gets up to 880 degrees Fahrenheit! That is hotter than fire!

And where there are volcanoes, there is lots and lots of bubbling LAVA!

Timbo likes to poke the lava with a stick. Luckily Tikki is prepared to put out any flames!

Rikki Asks: "What is lava made of?"

Lava is very, very hot and is made of mostly melted rocks and minerals.

THE SOLAR SYSTEM: Planet Venus

Earth is the third planet in our solar system. It has the perfect mix of air, water and sunlight for us to live.

EARTH

Floating beyond the clouds of Earth are more than two thousand satellites! Timbo loves to ride on those.

Tikki: Don't worry - the International Space Station astronauts can't see you. They are too far away. But they are learning lots of things while in outer space. They want to find out if people can live on other planets like Mars.

There is also the International Space Station, where astronauts from around the world do experiments.

Rikki Asks: "What do astronauts see when they look at Earth from space?"

Astronauts see Earth's blue oceans, green forests, and white clouds from space.

THE SOLAR SYSTEM: Planet Earth

The fourth planet in our solar system is Mars. It also has air, water and sunlight, so if we could live on Mars, we would be able to enjoy its seasons, including summer and winter.

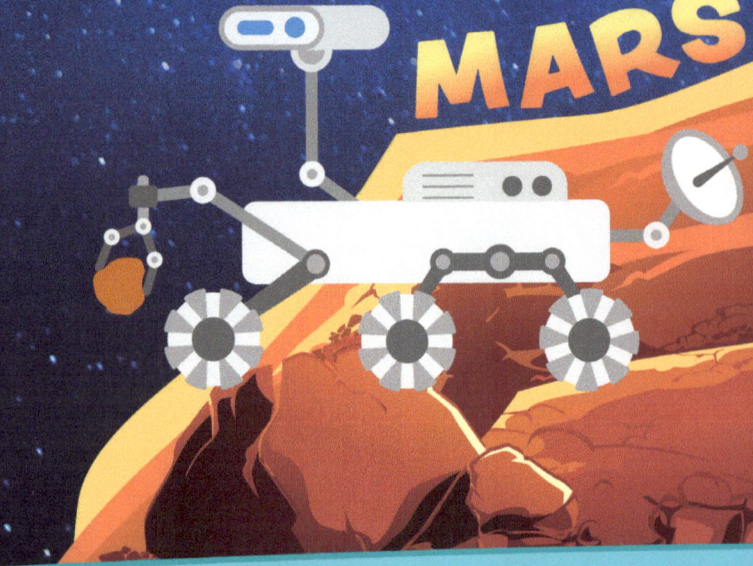

MARS

Rikki Asks: "Why is Mars called the Red Planet?"

Mars has lots of iron in its dirt, and this iron rusts, which gives the planet its unique red color.

Timbo: Mars is really fun to visit because I can make the best mud there! The mud castles I build on Mars look like I painted everything with dark red paint. Too bad I have to bring my own liquid water because all the water on Mars is a misty vapor or frozen underground.

Right now scientists at NASA are using the Curiosity Rover to explore Mars. They are testing to see what life would be like if we lived there.

Would you like to live on Mars one day?

THE SOLAR SYSTEM: Planet Mars

Jupiter is the fifth planet in our solar system and also the biggest. Jupiter is called a gas giant planet. A gas giant is a gigantic bubble filled with different gases swirling around a central core of metal, ice or rock.

Tikki: Between Mars and Jupiter is the Main Asteroid Belt. This giant ring contains trillions of large and small asteroid rocks, including the biggest asteroid which is called Ceres. The NASA Dawn Space Probe arrived at Ceres in March 2015. It took nearly eight years to get there.

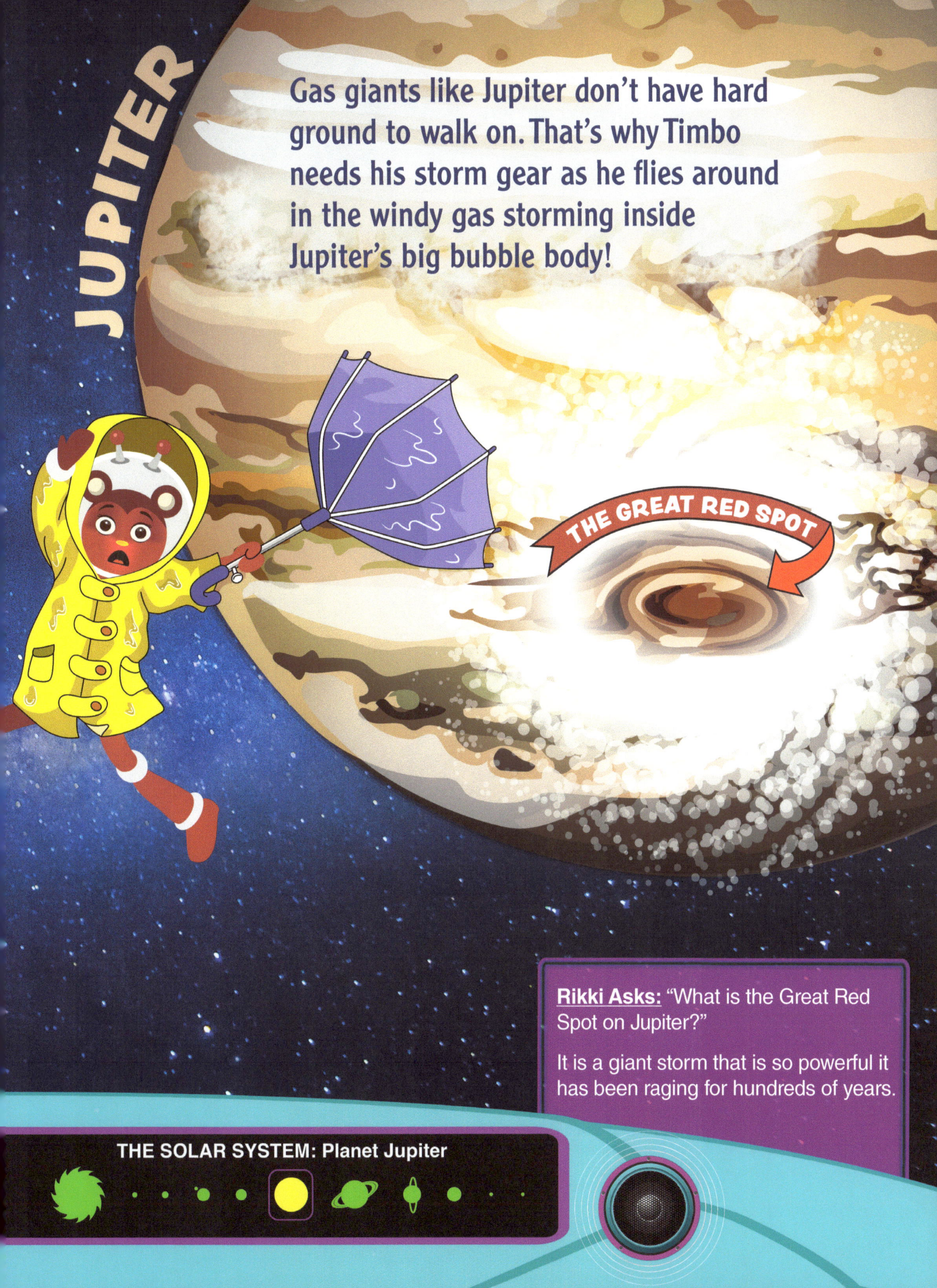

JUPITER

Gas giants like Jupiter don't have hard ground to walk on. That's why Timbo needs his storm gear as he flies around in the windy gas storming inside Jupiter's big bubble body!

THE GREAT RED SPOT

Rikki Asks: "What is the Great Red Spot on Jupiter?"

It is a giant storm that is so powerful it has been raging for hundreds of years.

THE SOLAR SYSTEM: Planet Jupiter

The sixth planet in the solar system is Saturn. It is also a gas giant planet like Jupiter and has awesome rings around it.

Timbo: Most people know Saturn by its cool rings, but very few people know that it takes Saturn thirty Earth years to make one complete orbit around the Sun. Earth's orbit around the Sun only takes us 365 days or one Earth year.

SATURN

From our home on Earth, the rings of Saturn look smooth enough to skate on.

Up close we see that Saturn's rings are made up of big chunks of ice and rock, as well as dust.

No way Timbo can skate on those rings!

Rikki Asks: "How long is a day on Saturn?"

Saturn rotates much faster than Earth, so a day on Saturn is less than half a day here on Earth.

THE SOLAR SYSTEM: Planet Saturn

The seventh planet in our solar system is Uranus. All the other planets spin like a top, but Uranus is tilted and spins on its side.

Tikki: Uranus sometimes gets very, very cold – dropping down to -224° Celsius (or -371° Fahrenheit) which is much colder than ice cream!

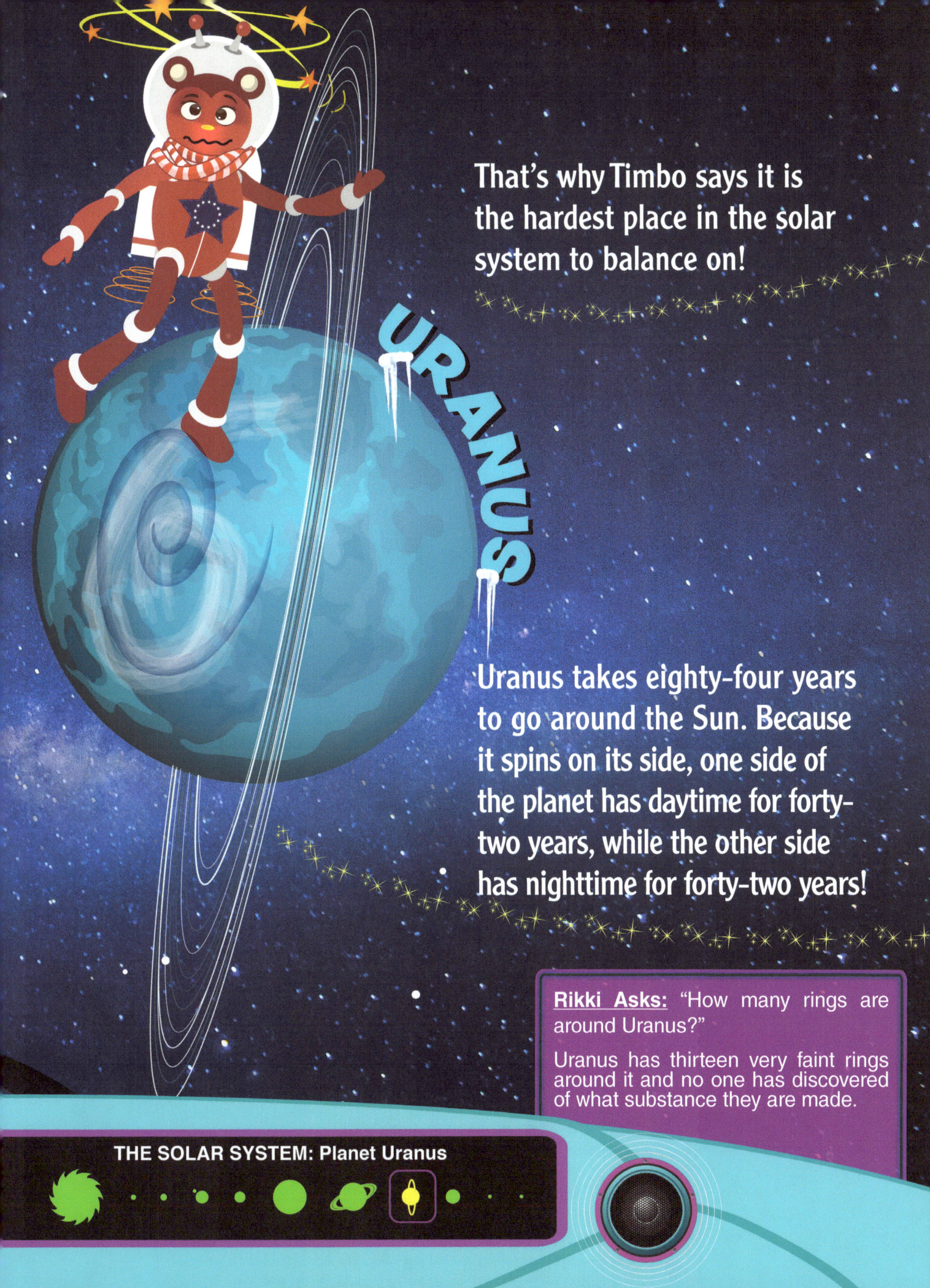

That's why Timbo says it is the hardest place in the solar system to balance on!

URANUS

Uranus takes eighty-four years to go around the Sun. Because it spins on its side, one side of the planet has daytime for forty-two years, while the other side has nighttime for forty-two years!

Rikki Asks: "How many rings are around Uranus?"

Uranus has thirteen very faint rings around it and no one has discovered of what substance they are made.

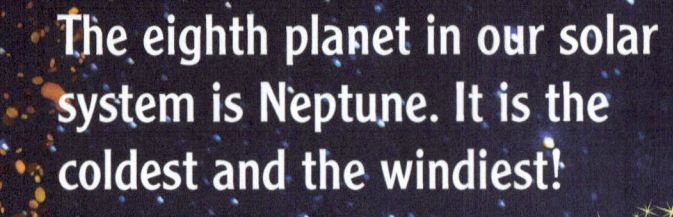

The eighth planet in our solar system is Neptune. It is the coldest and the windiest!

Rikki Asks: "How long does it take Neptune to orbit the Sun?"

It takes nearly 165 Earth years for Neptune to orbit the Sun, which is almost 164 more years than it takes our Earth.

Timbo: Neptune also has a giant storm like Jupiter. Neptune's giant storm is called the Great Dark Spot, which is about the size of Earth. Neptune's other storm, the Small Dark Spot, is about the size of Earth's moon.

NEPTUNE

Neptune has an average temperature of -214° Celsius (-353° Fahrenheit) which is too cold even for penguins!

Neptune also has high-speed solar winds that blow up to 832 miles per second. This is why Timbo loves to fly his kite there.

GREAT DARK SPOT

On the very far edge of our solar system are two small dwarf planets called Pluto and Eris.

PLUTO

Timbo: Eris is slightly bigger than Pluto and three times as far away from the Sun. Because Pluto's distance from the Sun is more than three billion miles, this means Timbo's golf ball would have to travel six billion miles or more to reach Eris.

These two planets are so far apart it would take hundreds and hundreds of years for Tikki's golf ball to travel between them.

ERIS

Rikki Asks: "Which one is bigger, Pluto or Eris?"

Eris is bigger than Pluto but both are smaller than our planet Earth.

THE SOLAR SYSTEM: Dwarf Planets Pluto and Eris

After visiting all the planets in the solar system, Rikki and her friends snuggle into a deep sleep.

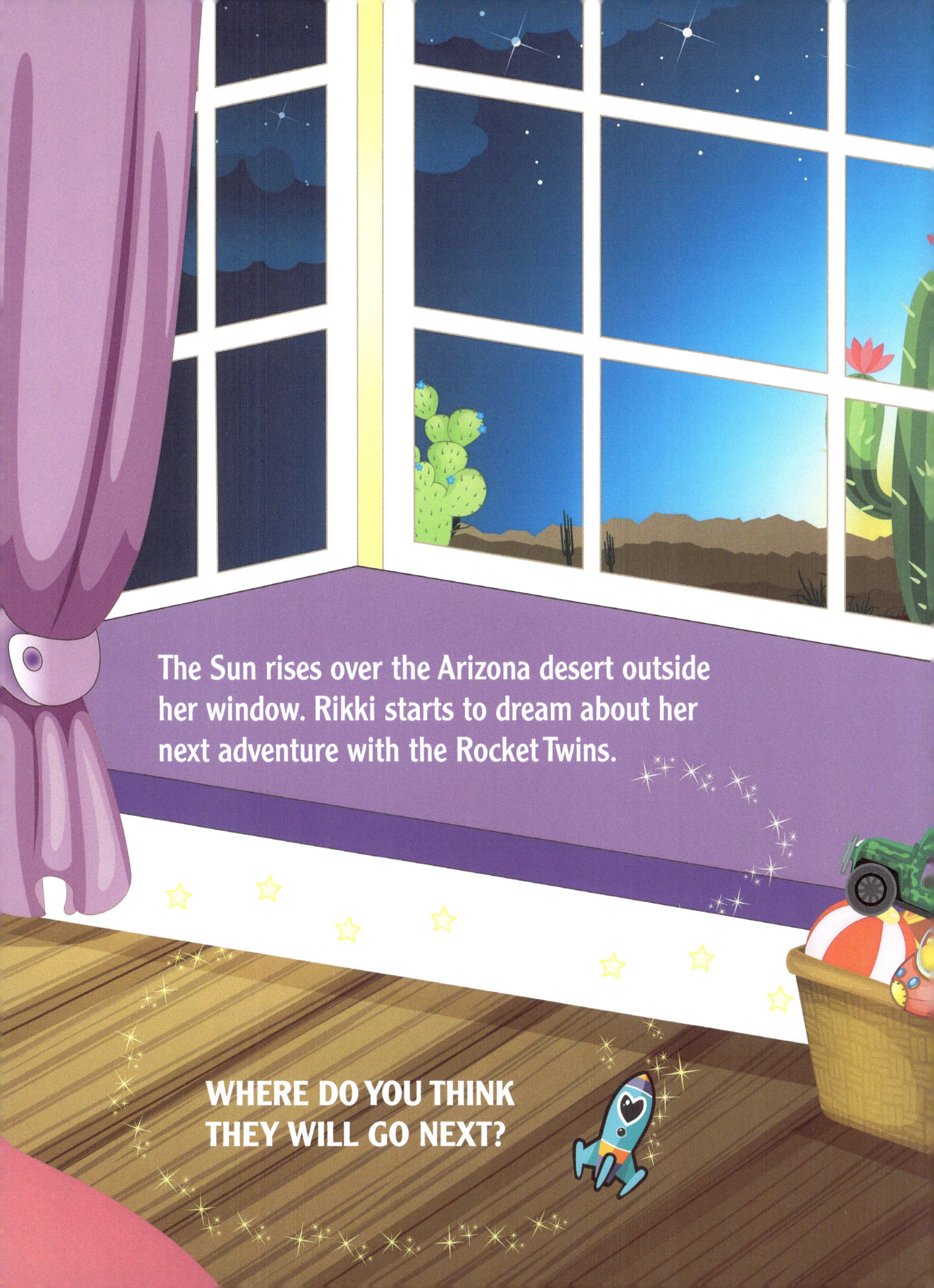

The Sun rises over the Arizona desert outside her window. Rikki starts to dream about her next adventure with the Rocket Twins.

WHERE DO YOU THINK THEY WILL GO NEXT?

AROUND THE SUN

Time traveling around the Sun

Distance from the Sun

PLUTO
- 248 years
- Between 3.67 and 4.4 billion miles

NEPTUNE
- 164.8 years
- 2.8 billion miles

URANUS
- 84 years
- 300 million miles

SATURN
- 29.5 years
- Between 850 to 940 million miles

JUPITER
- 11.9 years
- 467 million miles

MARS
- 687 Days
- 137 million miles

EARTH
- 365.25 Days
- 93 million miles

VENUS
- 224.7 Days
- 65 million miles

MERCURY
- 88 Days
- 35 million miles

THE PLANETS OF THE SOLAR SYSTEM

Outer space is a really big place! It is full of stars, comets, asteroids, planets, and galaxies and we're just one little spec in all of it. The first stop for the Discovery Rocket is within our own galaxy and the Solar System.

•**How big is the Solar System?** From the Sun all the way to the Kuiper Belt (where the dwarf planets of Pluto and Eres live), is about 7.5 billion kilometers or roughly 4.6 billion miles. But that's just to the Kuiper Belt; there's a lot past this area that scientists don't know about yet.

•**How long would it take to get there?** If you were to drive from the Sun to Pluto, it would take you 6,000 years. You should probably bring a few books or a couple of movies with you.

THE SUN

•**Does the Sun orbit?** Yes, it does! While planets go around the Sun, the Sun – actually, the entire Solar System – orbits around the center of the Milky Way.

•**How long does it take to orbit?** A really long time! It took 230 million years just to complete one orbit!

•**Sunshine Day:** We can't really calculate a day on the Sun. This is because the Sun does not rotate on an axis like a planet does. We calculate hours of a day by how the Sun returns to a certain spot in the sky.

IT'S HOT IN HERE! The Sun is classified as a yellow dwarf star. That means its surface temperatures are between 5027 and 5727 degrees Celsius.

MERCURY

•**How many Earths fit into Mercury?** None. Mercury is a very small planet, about a third the size of Earth. Actually, 3 Mercuries could fit inside the Earth!

•**How long does it take to orbit?** Being the closest planet to the Sun, it only takes 88 days for Mercury to orbit the Sun.

•**A Day on the Planet:** One day on the planet Mercury lasts as long as 59 days on Earth. That's the equivalent of a good part of our summer!

AND THE COW JUMPED OVER... Unlike some of the other planets, Mercury doesn't have any moons.

VENUS

•**How many Earths fit into Venus?** Earth is slightly bigger than Venus, so Venus could fit inside Earth. Because these two planets are somewhat similar, scientists call them sister planets.

•**A Day on the Planet:** It takes Venus 243 Earth days to rotate once on its axis. The planet's orbit around the Sun takes 225 Earth days, compared to the Earth's 365 so a day on Venus lasts longer than a year!

A VOLCANO A DAY: There are more than 1,600 volcanoes on Venus and scientists think there could be between 100,000 and 1 million more. Some are even still active!

EARTH

•**A Day on the Planet:** While the popular opinion is that there are 24 hours in a day, some scientists have stated that our days are actually only 23 hours long.

•**The Name Game:** Earth is the only planet that isn't named for a mythical God.

AND THE COW JUMPED OVER... We have one moon and her name is Luna. Scientists believe that Earth was originally two planets – the planet of Theia was knocked out of orbit and collided into Earth, merging the two together. This collision also created the moon. So you could think of the moon as the Earth's baby.

MARS

•**How many Earths fit into Mars?** Mars, like Venus, would be able to fit inside of Earth as it's slightly smaller.

•**A Day on the Planet:** Like our sister planet, Venus, Mars is remarkably similar to Earth. A day on Mars is about 39 minutes longer than our standard 24 hours.

I'M CURIOUSLY MARS: Curiosity, the rover launched by NASA, is currently on Mars trying to answer a very important question – could Mars ever have supported small life forms? The answer to this could help scientists see if Mars is another inhabitable planet, like Earth.

JUPITER

•**How many Earths fit into Jupiter?** Jupiter is so big, 1,300 Earths could fit inside of it.

•**A Day on the Planet:** A day only lasts a little over 9 hours, about the time you're in school or your parents are at work. Jupiter has some of the shortest days in the Solar System.

AND THE COW JUMPED OVER... So many moons! Currently, we know of 67 moons from Jupiter, with Ganymede being the largest.

SATURN

•**How many Earths fit into Saturn?** Saturn is nearly 10 times wider than Earth, so about 750 Earths would fit inside.

•**A Day on the Planet:** A day on Saturn is only about 10 hours, just about an hour longer than a day on Jupiter and 14 hours shorter than what we experience on Earth.

THE NAME GAME: Saturn gets its name from the Roman god of agriculture and time. If Saturn is your favorite planet, throw a party on a Saturday – it's also named after the Roman god.

URANUS

•**How many Earths fit into Uranus:** You could fit 63 Earths into Uranus. As the third largest planet, it's four times the size of our planet.

•**The Name Game:** George was going to be the original name of the planet, as proposed by discoverer William Herschel. The science community didn't really like it, so they named it Ouranos or Uranus, after the Greek God of the sky.

TILT-A-WHIRL: Uranus is the only planet that is tilted on its side.

NEPTUNE

•**How many Earths fit into Neptune?** You could fit 58 Earths within the planet of Neptune.

•**A Day on the Planet:** Like the other gas giant planets, Neptune's days are very short. They only last for 16 hours.

BABY, IT'S COLD OUTSIDE: Uranus may be the coldest planet in the Solar System, but Neptune's moon, Triton, is the coldest place in the Solar System, with surface temperatures of -235 degrees Celsius or -391 degrees Fahrenheit.

PLUTO

•**How many Earths fit into Pluto?** Pluto is actually a very small planet, even smaller than Mercury. You could fit 151 Plutos into Earth.

•**Demoted to Extra:** In 2006, scientists reclassified Pluto from its status as our ninth planet to that of a dwarf planet. This was because scientists formalized the definition of a planet, which Pluto no longer met.

THE PLUTO ZONE: Even though Pluto is now considered a dwarf planet, it has helped scientists discover a third zone within the Solar System – the Kuiper Belt.

UNDERSTANDING YOUR WORLD

Sometimes on learning adventures we come across big words. Big words are easy to master if you know what they mean. Which one of these words is going to become your favorite?

ASTRONAUT EASY SPACE DEFINITIONS FOR THE SOLAR SYSTEM & BEYOND

Satellite – A moon, planet, or machine that orbits around another planet or a star.

Space Station – A spacecraft that can support a crew so they can remain in space for an extended amount of time.

Moon – Our moon is the only permanent natural satellite. We have learned many things about the moon by traveling to it in a spacecraft.

Planet – An object that moves around a star. Planets are also known as satellites.

Astronaut – A person who is trained to serve aboard a spacecraft.

Outer Space – The area above the atmosphere of Earth, where our solar system, comets, asteroids, and other objects live.

Solar System – The solar system is where our planet Earth lives, along with comets, asteroids, meteoroids, and more.

Star – Any heavenly body, except the moon, that appear as fixed bright lights in the night sky.

Orbit – A circular path that an object makes in space.

Gravity – The force of attraction by which terrestrial bodies tend to fall toward the center of the earth.

Spacesuit – Worn by an astronaut so they can move around in outer space.

Volcano – An opening in the earth's crust that can erupt and send out lava, steam, ashes, etc.

Asteroid Belt – A circular area between the planets of Mars and Jupiter, made up of asteroids and minor planets.

Gas Planet – A planet that is made up of different gases, such as hydrogen and helium.

Terrestrial Planets – An inner planet, made up of rocks or metals and has a solid surface.

Dwarf Planets – A spherical celestial body revolving around the sun, similar to a planet, but not large enough to gravitationally clear its orbital region.

Continue the adventure online at
www.discoveryrocket.com!

Join the Discovery Rocket Club for exclusive content
and free learning resources, including a graphic
overview of the Solar System.

Visit our online shop or learn more about Rikki's
upcoming adventures in math and science.

Don't forget to download the Discovery Rocket
mobile game for Android and iOS.
discoveryrocket.com/shop
It includes interactive learning and fun play
levels for kids 3 and older. (Even your savvy
8-year-old will think it's fun!)